BLACK HOLE

EXPLORING THE MYSTERIES

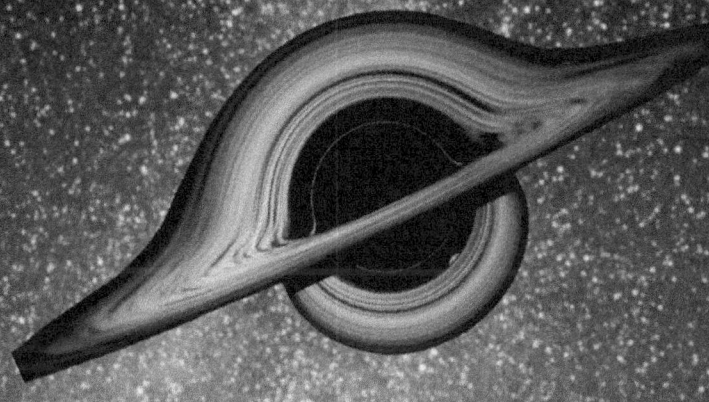

by Amazir Amanath (Dip. in Celestial Navigation)

Exploring the Mysteries of Black Holes: The Mystery Spot

By:
Amazir Amanath
(Aspiring Astronaut
Dip. in Celestial Navigation)

Words of Writer

Embarking on the exhilarating journey of adolescence, at a mere 15 years old, I find myself at the intersection of dreams and discovery. An unwavering passion for the cosmos propels me forward as an aspiring astronaut and devoted physics enthusiast. In the tapestry of my young life, I proudly weave the narrative of "Exploring the Mysteries of Black Holes" a testament to my insatiable curiosity and eagerness to unravel the profound mysteries of the universe.

Despite my youth, the pages of this book bear witness to my earnest commitment to understanding the complexities of physics. With each word, I navigate the intricacies of Einstein's revolutionary theories, translating the esoteric language of relativity into a narrative that transcends age barriers. It is a labor of love and a testament to my determination to bridge the gap between the profound and the accessible.

Crafting "Exploring the Mysteries of Black Holes" has been a journey of self-discovery, an odyssey into the heart of theoretical physics that has both challenged and inspired me. The ambition to comprehend the implications of Einstein's theories has not only fueled my intellectual growth but has also ignited a fervent desire to contribute to the discourse surrounding the cosmos.

In the constellation of my aspirations, the dream of becoming an astronaut shines brightly. The yearning to explore the cosmos, to transcend earthly boundaries, drives me to pursue knowledge with unwavering dedication. My story is one of a young mind reaching for the stars, fueled by an unquenchable thirst for understanding and a determination to leave an indelible mark on the scientific landscape.

As I pen my narrative, I aim to inspire others to embrace the boundless possibilities that lie within the pursuit of knowledge. In the vast expanse of the universe, age is but a fleeting factor and the journey of self-discovery knows no bounds.

Amazir Amanath
Writer - Exploring the Mysteries of Black Holes

Exploring the Mysteries of Black Holes | Amazir Amanath | 1st Edition

Chapter 1

Exploring the Mysteries of Black Holes | Amazir Amanath | 1st Edition

Introduction to Black Holes: Unraveling the Cosmic Miracle

1. Understanding Black Holes

Black holes, enigmatic and mysterious, are celestial phenomena that captivate the imagination and challenge our understanding of the universe. At their core lies a concept so profound and paradoxical that it defies conventional wisdom: gravitational collapse. Imagine a region of space where gravity is so intense that nothing, not even light, can escape its grasp. This is the essence of a black hole.

Gravitational collapse occurs when a massive star reaches the end of its life cycle. Throughout its existence, a star generates energy through nuclear fusion, counteracting the inward pull of gravity with the outward pressure of radiation. However, once the star exhausts its nuclear fuel, this delicate balance collapses. Without the energy to resist gravity, the star's core undergoes a rapid and catastrophic collapse, leading to the formation of a black hole.

2. History and Evolution of Black Holes

The concept of black holes traces its roots back to the early 20th century, with Albert Einstein's revolutionary theory of general relativity. Einstein's equations provided the theoretical framework for understanding how gravity warps the fabric of spacetime. However, it wasn't until the mid-20th century that scientists began to seriously consider the possibility of black holes.

In 1916, Karl Schwarzschild, a German physicist, derived the first exact solution to Einstein's equations, describing a black hole's gravitational field. This laid the groundwork for further exploration into these enigmatic objects. However, it wasn't until the 1960s that the term "black hole" was coined by physicist John

Archibald Wheeler, encapsulating the idea of an object from which nothing can escape.

The study of black holes took a giant leap forward with the advent of space-based observatories and advanced telescopes. Beginning in the late 20th century and continuing into the 21st century, astronomers have made groundbreaking discoveries that have deepened our understanding of black holes. From the detection of stellar-mass black holes in binary systems to the observation of supermassive black holes at the centers of galaxies, these findings have reshaped our view of the cosmos.

3. Modern Observations and Theoretical Advancements

Modern observations have revealed a diverse array of black holes, ranging from stellar-mass black holes to supermassive behemoths millions to billions of times the mass of our Sun. These observations have provided valuable insights into the formation and evolution of black holes, shedding light on their role in shaping the structure of galaxies and the cosmos at large.

Furthermore, advancements in theoretical physics have deepened our understanding of black holes and their behavior. Concepts such as Hawking radiation, proposed by Stephen Hawking in 1974, suggest that black holes aren't entirely black; they emit radiation due to quantum effects near the event horizon.

Exploring the Mysteries of Black Holes | Amazir Amanath | 1st Edition

This revelation challenged long-held assumptions about the nature of black holes and opened new avenues for exploration.

In recent years, the field of black hole research has entered a golden age, fueled by technological advancements and interdisciplinary collaboration. From the detection of gravitational waves produced by black hole mergers to the imaging of the supermassive black hole at the center of the galaxy M87, these breakthroughs have ushered in a new era of discovery, pushing the boundaries of our knowledge and inspiring future generations to unravel the mysteries of the cosmos.

Finally, Black holes, with their gravitational might and inscrutable nature, stand as one of the most captivating phenomena in the universe. From their theoretical origins in Einstein's equations to their modern-day detection and study, the journey to understanding black holes has been one of profound discovery and intellectual exploration. As we continue to peer into the depths of space and probe the secrets of the cosmos, black holes will undoubtedly remain a central focus of scientific inquiry, offering glimpses into the fundamental nature of reality and our place within it.

Chapter 2

Formation of Black Holes

Black holes are some of the most mysterious objects in the universe, exerting an immense gravitational pull from which not even light can escape. They can form through various processes, the most common being the collapse of massive stars and the mergers of compact objects such as neutron stars or other black holes.

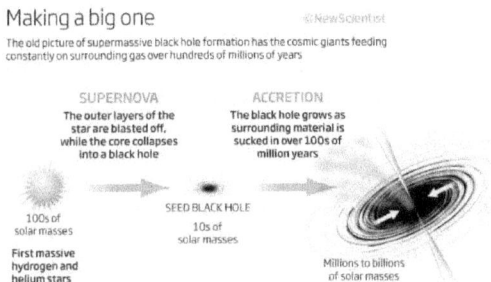

Formation from the Collapse of Massive Stars:

One of the most well-understood mechanisms for black hole formation is through the collapse of massive stars. When a massive star reaches the end of its life cycle, it undergoes a catastrophic event known as a supernova explosion. During a supernova, the outer layers of the star are expelled into space, leaving behind a dense core.

If the remaining core of the star has a mass greater than about three times that of the Sun (the Tolman-Oppenheimer-Volkoff limit), gravity overwhelms the internal pressure exerted by the matter, causing the core to collapse under its own weight. This collapse continues until all the mass is concentrated into a single point of infinite density, known as a singularity, surrounded by an event horizon - the boundary beyond which nothing can escape.

This process results in the formation of a stellar-mass black hole, with a mass ranging from a few times that of the Sun to several tens of times the solar mass.

Formation through Compact Object Mergers:

Black holes can also form through the merger of compact objects such as neutron stars or other black holes. When two such objects orbit each other closely, they emit gravitational waves, which carry away energy and angular momentum. As a result, the objects spiral inward until they eventually merge into a single, more massive black hole. This process of compact object mergers has been observed directly through gravitational wave detections by instruments such as LIGO (the Laser Interferometer Gravitational-Wave Observatory) and Virgo. These mergers can lead to the formation of black holes with masses larger than those produced by the collapse of individual stars.

Types of Black Holes:

Black holes can be categorized based on their mass and size, resulting in different types such as stellar-mass black holes and supermassive black holes.

1. Stellar-Mass Black Holes

Stellar-mass black holes are typically formed from the collapse of massive stars, as described earlier. They have masses ranging from a few times that of the Sun to several tens of solar masses. These black holes are relatively common in the universe and are often found in binary systems with companion stars.

Stellar-mass black holes are characterized by their relatively small size and high density. They exert a strong gravitational pull on nearby objects, which can distort the spacetime around them and cause visible effects such as gravitational lensing.

2. Intermediate-Mass Black Holes

Intermediate-mass black holes are believed to have masses between stellar-mass black holes and supermassive black holes, ranging from hundreds to thousands of solar masses. These black holes are more elusive and have been challenging to detect due to their smaller size and lower luminosity compared to supermassive black holes. Intermediate-mass black holes may form through the merger of stellar-mass black holes in dense star clusters or through the accretion of gas and stars in galactic centers. Their existence is supported by indirect observational evidence, such as the presence of dense star clusters and unusual X-ray emissions in some galaxies.

3. Supermassive Black Holes

Supermassive black holes are the largest and most massive type of black holes known to exist, with masses ranging from millions to billions of times that of the Sun. These black holes reside at the centers of most galaxies, including our own Milky Way.

The exact mechanisms of supermassive black hole formation are still a topic of active research, but they are believed to grow through a combination of accretion of surrounding matter and mergers with other black holes. Supermassive black holes are often surrounded by accretion disks of hot gas and dust, which emit intense radiation across the electromagnetic spectrum.

Supermassive black holes play a crucial role in the evolution of galaxies, influencing their structure, dynamics, and star formation rates. They are also thought to be responsible for powering the energetic phenomena observed in active galactic nuclei, such as quasars and blazars.

Conclusion

Black holes are fascinating objects that arise from the extreme conditions of gravity and density in the universe. They can form through the collapse of massive stars or the mergers of compact objects, giving rise to various types such as stellar-mass, intermediate-mass, and supermassive black holes. Understanding the formation and properties of black holes is essential for unraveling the mysteries of the cosmos and advancing our knowledge of fundamental physics.

Chapter 3

Exploring the Mysteries of Black Holes | Amazir Amanath | 1st Edition

Characteristics of Black Holes

Introduction
Black holes are one of the most fascinating and enigmatic phenomena in the universe. They are regions in space where the gravitational pull is so intense that nothing, not even light, can escape from them. Understanding the key properties of black holes, such as their event horizons, singularities, and gravitational effects, is crucial for unraveling the mysteries of these cosmic giants.

1. Event Horizons
At the heart of every black hole lies its event horizon, a boundary beyond which escape is impossible. This boundary marks the point of no return for anything venturing too close to the black hole. Once an object crosses the event horizon, it is inexorably drawn towards the singularity at the center.

The event horizon's size is directly related to the mass of the black hole. For a non-rotating (Schwarzschild) black hole, the event horizon is a spherical surface. However, for rotating black holes (Kerr black holes), the event horizon takes on a more complex shape, known as the ergosphere, where space itself is dragged along with the black hole's rotation.

2. Singularities
At the center of a black hole lies the singularity, a point of infinite density where the laws of physics, as we currently understand them, break down. Singularities are hidden behind the event horizon, making them inaccessible to observation. According to general relativity, the singularity represents a breakdown in spacetime curvature, where the gravitational forces become infinitely strong.

3. Gravitational Effects
The gravitational effects of black holes are profound and far-reaching. Near a black hole, the gravitational field is so intense that it distorts the paths of light and matter. This phenomenon, known as gravitational lensing, can bend light rays around the black hole, creating distorted images of objects behind it.

Black holes also exhibit time dilation effects. Clocks near a black hole appear to run slower compared to those further away, due to the immense gravitational pull slowing down the passage of time.

4. Interaction with Matter and Light

Black holes interact with matter and light in various ways, shaping the surrounding environment in dramatic fashion. When matter falls towards a black hole, it forms an accretion disk—a swirling disk of gas and dust spiraling towards the event horizon. As the matter in the accretion disk accelerates, it releases vast amounts of energy in the form of X-rays and other electromagnetic radiation.

Jets are another fascinating phenomenon associated with black holes. These narrow streams of high-energy particles are ejected from the vicinity of black holes at near-light speeds. Jets are thought to originate from the intense magnetic fields generated near the black hole's event horizon, which accelerate particles to relativistic speeds along the axis of rotation.

5. Observational Evidence

While black holes themselves cannot be directly observed, their presence can be inferred through the effects they have on their surroundings. Astronomers use various techniques, such as studying the motion of stars and gas clouds near the center of galaxies, to detect the gravitational influence of black holes. Additionally, the emission of X-rays from accretion disks and the detection of jets provide further evidence for the existence of black holes.

6. The Role of Black Holes in Galactic Evolution

Black holes play a crucial role in the evolution of galaxies. Supermassive black holes, which reside at the centers of most galaxies, regulate the growth of their host galaxies by influencing the formation of stars and the distribution of matter. The energetic processes associated with black holes, such as the release of radiation from accretion disks and jets, can heat up and ionize the surrounding gas, affecting the formation of new stars.

Conclusion

Black holes are extraordinary cosmic phenomena that challenge our understanding of the universe. Their unique properties, such as event horizons, singularities, and gravitational effects, make them both fascinating and mysterious objects of study. By unraveling the mysteries of black holes, scientists hope to gain deeper insights into the fundamental laws of physics and the evolution of the cosmos.

Chapter 4

Observational Evidence for the Existence of Black Holes

Abstract:
Black holes, enigmatic cosmic entities predicted by Einstein's theory of general relativity, have been a subject of fascination and intense study for decades. While the idea of black holes was initially met with skepticism, observational evidence collected over the years has solidified their existence. This paper reviews the observational evidence supporting the existence of black holes, encompassing studies of stellar dynamics, gravitational lensing phenomena, and the recent detection of gravitational waves.

1. Introduction:
The concept of black holes, regions of spacetime where gravity is so intense that not even light can escape, was first proposed by physicist Karl Schwarzschild in 1916. Initially met with skepticism, black holes gradually gained acceptance as a consequence of theoretical developments and observational evidence. This paper explores the various observational techniques used to study black holes and the compelling evidence supporting their existence.

2. Stellar Dynamics:

One of the primary methods used to detect black holes is through the study of stellar dynamics within galactic centers. Observations of stars orbiting around invisible companions with extremely high masses provide compelling evidence for the presence of black holes. By tracking the motion of these stars over time, astronomers can infer the presence of a massive, compact object that exerts gravitational influence but emits no detectable light. Notable examples include the supermassive black hole Sagittarius A* at the center of our Milky Way galaxy and the black hole in the galaxy M87, imaged for the first time in 2019 by the Event Horizon Telescope.

3. Gravitational Lensing:

Another significant piece of evidence for black holes comes from the phenomenon of gravitational lensing. According to general relativity, massive objects can bend the path of light rays passing near them, causing distortions in the images of background objects. When a black hole passes between Earth and a distant light source, it can act as a gravitational lens, producing characteristic magnification and distortion patterns. By observing these effects, astronomers can infer the presence and properties of intervening black holes. Gravitational lensing has been instrumental in detecting both stellar-mass and supermassive black holes, providing crucial insights into their distribution and demographics across the universe.

4. Detection of Gravitational Waves:

In 2015, the Laser Interferometer Gravitational-Wave Observatory (LIGO) made history by detecting gravitational waves for the first time, confirming a key prediction of Einstein's theory of general relativity. Gravitational waves are ripples in spacetime caused by the acceleration of massive objects, such as merging black holes or neutron stars. Since then, LIGO and its European counterpart Virgo have made several detections of binary black hole mergers, providing direct observational evidence for the existence of black holes. These detections not only confirm the existence of black holes but also offer insights into their masses, spins, and merger rates, furthering our understanding of these enigmatic objects.

5. Conclusion:

Observational evidence collected over the years has firmly established the existence of black holes, from the study of stellar dynamics and gravitational lensing to the recent detection of gravitational waves. These cosmic entities, once regarded as purely theoretical constructs, are now recognized as fundamental components of the universe. Continued observations and technological advancements promise to unveil further insights into the nature and behavior of black holes, opening new avenues for exploration in astrophysics and cosmology.

Exploring the Mysteries of Black Holes | Amazir Amanath | 1st Edition

Chapter 5

The Role of Black Holes in Astrophysical Phenomena

Introduction
Astrophysical phenomena captivate our imagination with their sheer scale and complexity. Among the most enigmatic entities in the cosmos are black holes, gravitational beasts whose influence extends far beyond their event horizons. In this discourse, we embark on a journey to explore the multifaceted roles of black holes in shaping various astrophysical phenomena, from the grand orchestration of galaxy formation and evolution to the intricate dance of star formation and the dynamics of galactic nuclei.

Galaxy Formation and Evolution
Galaxies, the building blocks of the universe, owe their structures and behaviors to the interplay of various cosmic forces, among which black holes play a pivotal role. At the heart of most, if not all, massive galaxies lie supermassive black holes (SMBHs). These cosmic behemoths, with masses ranging from millions to billions of times that of our Sun, wield immense gravitational influence, sculpting the very fabric of their host galaxies.

During the epoch of galaxy formation, the presence of SMBHs is believed to have played a crucial role in regulating star formation rates and shaping the properties of galaxies. The process begins with the collapse of primordial gas clouds under gravity, leading to the formation of protogalactic structures. As gas accretes onto the central SMBH, it releases copious amounts of energy in the form of radiation and powerful outflows, known as active galactic nuclei (AGN) feedback.

AGN feedback mechanisms, fueled by accretion onto SMBHs, can profoundly impact the surrounding gas reservoirs, regulating star formation by heating or expelling gas from galaxies. This feedback loop acts as a cosmic thermostat, balancing the inward collapse of gas with the outward expulsion of material, thereby regulating the growth of galaxies over cosmic time.

Furthermore, observations suggest a close relationship between SMBHs and their host galaxies, known as the black hole-galaxy coevolution paradigm. This intimate connection implies that the growth of SMBHs and the assembly of galaxies are intricately linked processes, with each influencing the evolution of the other over cosmic epochs.

Star Formation

In the cosmic theater of star formation, black holes cast a long shadow, influencing the birth, life, and death of stars across the universe. Stellar nurseries, regions of dense molecular clouds where stars are born, are often found in the vicinity of massive black holes, such as those lurking at the centers of galaxies.

The gravitational influence of SMBHs can trigger the collapse of nearby gas clouds, compressing them to the point of gravitational instability. As these clouds collapse, they fragment into smaller clumps, each destined to become a new generation of stars. The presence of a central black hole can enhance this process through tidal forces, stirring up the surrounding gas and promoting the formation of massive star clusters.

However, the relationship between black holes and star formation is not solely constructive. In some cases, the intense radiation and energetic outflows from accreting black holes can hinder star formation by dispersing or heating the surrounding gas, thereby suppressing the birth of new stars. This dichotomy underscores the complex interplay between black hole activity and the formation of stellar populations within galaxies.

Galactic Nuclei Dynamics

At the heart of many galaxies lie compact regions of extraordinary density and activity known as galactic nuclei. These dense hubs, which often host SMBHs, are the gravitational fulcrum around which the dynamics of entire galaxies revolve.

The presence of a central black hole can profoundly influence the dynamics of galactic nuclei through a variety of mechanisms. In spiral galaxies, such as our Milky Way, the motion of stars and gas in the central region is shaped by the gravitational potential of the SMBH. Observations of stars orbiting the Milky Way's central SMBH, Sagittarius A*, have provided compelling evidence for the existence of supermassive black holes and offered insights into their mass and properties.

In elliptical galaxies, which lack the spiral structure of their counterparts, the dynamics of galactic nuclei are thought to be dominated by the gravitational interactions between stars and the central SMBH. These interactions can lead to phenomena such as stellar tidal disruptions, where stars stray too close to the black hole and are torn apart by its immense gravitational pull.

Furthermore, the presence of SMBHs can have profound implications for the long-term evolution and fate of galactic nuclei. In some cases, the accretion of gas onto the central black hole can trigger energetic outbursts known as galactic mergers, leading to the formation of quasars or active galactic nuclei (AGN). These cosmic fireworks inject vast amounts of energy into the surrounding environment, driving galactic-scale winds and shaping the evolution of entire galaxy clusters.

Conclusion

In the tapestry of the cosmos, black holes emerge as cosmic architects, sculpting the landscape of galaxies and shaping the destiny of stars. From the grand sweep of galaxy formation and evolution to the delicate dance of star birth and the dynamics of galactic nuclei, black holes exert a profound influence on the fabric of the universe. As we continue to probe the mysteries of the cosmos, the role of black holes in astrophysical phenomena remains a fertile ground for exploration, offering insights into the fundamental forces that govern the cosmos.

Chapter 6

The Role of Black Holes in Astrophysical Phenomena

Introduction

Black holes, the enigmatic cosmic phenomena predicted by Einstein's theory of general relativity, have captivated the imaginations of scientists and the public alike for decades. These celestial entities, characterized by their immense gravitational pull, have puzzled astronomers with their mysterious properties and unconventional behavior. In this exploration, we delve into the intricate anatomy of black holes, from their exterior structure to the perplexing depths of their interiors. We will unravel concepts such as the "no-hair" theorem and the information paradox, shedding light on the fundamental principles that govern these cosmic giants.

Exterior Structure

At first glance, a black hole appears as a region of space from which nothing, not even light, can escape. This boundary, known as the event horizon, marks the point of no return for any object venturing too close to the black hole's gravitational grasp. Surrounding the event horizon is the accretion disk, a swirling mass of gas, dust, and other celestial debris drawn in by the black hole's immense gravitational pull. The accretion disk emits intense radiation as it spirals inward, providing astronomers with valuable insights into the properties of black holes.

Interior Geometry

Beyond the event horizon lies the heart of the black hole: the singularity. According to general relativity, the singularity is a point of infinite density and zero volume, where the laws of physics, as we understand them, break down. Surrounding the singularity is the region known as the black hole's interior, which is shrouded in mystery due to the extreme conditions within. The exact geometry of the interior remains a subject of debate among physicists, with theories ranging from a point-like singularity to more exotic possibilities such as a wormhole or a structure known as a Planck star.

The No-Hair Theorem

One of the most intriguing aspects of black holes is captured by the "no-hair" theorem, which suggests that black holes possess only three measurable properties: mass, electric charge, and angular momentum. According to this theorem, all other information about the matter that formed the black hole is lost once it crosses the event horizon, leading to the notion that black holes have "no hair" – they are characterized by their simplicity rather than the complexity of their constituents. While the no-hair theorem has been supported by numerous observations and theoretical models, its implications raise profound questions about the nature of information and the fundamental principles of physics.

The Information Paradox

The concept of the information paradox arises from the apparent conflict between the predictions of general relativity and the principles of quantum mechanics. According to quantum mechanics, information is always conserved, meaning that any process that destroys information violates the fundamental laws of physics. However, in the case of black holes, the no-hair theorem suggests that information about the matter that falls into a black hole is irretrievably lost, leading to a violation of quantum mechanics. This paradox has sparked intense debate among physicists and has prompted the exploration of theories such as black hole complementarity and the holographic principle in an attempt to reconcile the conflicting principles of general relativity and quantum mechanics.

Emerging Insights and Future Directions

Despite the many mysteries that surround black holes, recent advancements in observational techniques and theoretical models have provided new insights into these cosmic enigmas. Observatories such as the Event Horizon Telescope have enabled astronomers to directly image the silhouette of a black hole's event horizon for the first time, providing valuable data for testing theoretical predictions. Meanwhile, developments in quantum gravity and string theory offer promising avenues for resolving the information paradox and unlocking the secrets of black hole physics.

Conclusion

In the vast tapestry of the cosmos, black holes stand as some of the most enigmatic and fascinating entities. From their exterior structure to the depths of their singularities, these cosmic behemoths challenge our understanding of the universe and push the boundaries of theoretical physics. As we continue to probe the mysteries of black holes, we embark on a journey of discovery that promises to unveil the fundamental truths that govern the cosmos and our place within it.

Chapter 7

Black Hole Evolution

Introduction

Black holes are some of the most enigmatic and fascinating objects in the universe. Formed from the remnants of massive stars, they possess gravitational pulls so strong that not even light can escape from them, making them invisible to direct observation. The life cycle of a black hole spans billions of years, involving various stages of evolution and eventual demise. In this paper, we will delve into the formation of black holes, their evolution over time, and the processes that lead to their eventual fate, including Hawking radiation and evaporation.

Formation of Black Holes

Black holes form as a result of the gravitational collapse of massive stars. When a star exhausts its nuclear fuel, it can no longer sustain the outward pressure generated by nuclear fusion in its core. Without this pressure to counteract gravity, the star begins to collapse under its own weight. For massive stars, this collapse can lead to a catastrophic event known as a supernova.

During a supernova explosion, the outer layers of the star are ejected into space, leaving behind a dense core. If the core's mass exceeds a critical threshold known as the Chandrasekhar limit (about 1.4 times the mass of the Sun), gravity overwhelms all other forces, causing the core to collapse further. This collapse continues until the core is compressed into an infinitely dense point known as a singularity, surrounded by an event horizon—the boundary beyond which nothing can escape.

Evolution of Black Holes

Once formed, black holes can evolve through various mechanisms, primarily through the accretion of matter and the merging with other black holes. Accretion occurs when matter from the surrounding space falls into the gravitational field of the black hole. As this matter spirals inward, it forms an accretion disk around the black hole, generating intense radiation across the electromagnetic spectrum. This process can release vast amounts of energy and is responsible for phenomena such as quasars and active galactic nuclei (AGNs).

Black holes can also grow in size through mergers with other black holes. When two black holes come into close proximity, they can spiral inward due to gravitational waves, eventually merging to form a larger black hole. This process has been observed indirectly through the detection of gravitational waves by instruments such as LIGO (the Laser Interferometer Gravitational-Wave Observatory).

Hawking Radiation and Evaporation

While black holes are known for their immense gravitational pull, they are not completely dark. According to quantum mechanics, particles and antiparticles are constantly being created and annihilated near the event horizon of a black hole. Occasionally, one of these particle-antiparticle pairs may appear just outside the event horizon, with one particle falling into the black hole while the other escapes into space. This process, predicted by physicist Stephen Hawking, results in the emission of radiation from the black hole, known as Hawking radiation. Over time, this radiation causes the black hole to lose mass and energy, leading to its eventual evaporation. For small black holes, this process is negligible, as they emit very little Hawking radiation. However, for larger black holes, such as those formed from stellar collapse, Hawking radiation becomes more significant over time.

As a black hole emits Hawking radiation, its mass decreases, causing its gravitational pull to weaken. Eventually, the rate of radiation emission exceeds the rate of matter accretion, leading to a runaway process where the black hole loses mass rapidly. As the black hole evaporates, it releases a burst of energy in its final moments, known as a black hole explosion or gamma-ray burst.

The Fate of Black Holes

The ultimate fate of a black hole depends on its mass. For stellar-mass black holes formed from the collapse of massive stars, the process of evaporation through Hawking radiation will eventually cause them to disappear entirely. This process, however, takes an incredibly long time—far longer than the current age of the universe for most black holes.

For supermassive black holes found at the centers of galaxies, the situation is different. These black holes can have masses millions or even billions of times that of the Sun and are thought to play a crucial role in the evolution of galaxies. While they also emit Hawking radiation, the surrounding environment provides ample opportunities for matter accretion, balancing out the loss of mass due to radiation. As a result, supermassive black holes are expected to persist for vast periods of time, potentially outliving the stars in their host galaxies.

Conclusion

The life cycle of a black hole is a complex and fascinating journey that spans billions of years. From their formation through the collapse of massive stars to their eventual evaporation through Hawking radiation, black holes represent some of the most extreme and mysterious objects in the universe. While their ultimate fate may vary depending on their mass, one thing is clear: black holes will continue to capture the imagination of scientists and astronomers for generations to come, shedding light on the nature of space, time, and gravity.

Chapter 8

Exploring Extreme Environments Around Black Holes

Introduction

Black holes, the enigmatic cosmic entities formed from the collapse of massive stars, harbor some of the most extreme environments in the universe. Within their vicinity, gravity reaches staggering levels, magnetic fields become immensely powerful, and particles attain velocities close to the speed of light. In this comprehensive exploration, we delve into the mesmerizing realms surrounding black holes, uncovering the profound effects they exert on their surroundings and the fundamental processes that shape these extreme environments.

The Gravity of the Situation

At the heart of a black hole lies a gravitational singularity, a point where the density becomes infinite and the laws of physics as we know them break down. The gravitational pull exerted by this singularity is so immense that it warps the fabric of space-time itself, creating what is known as a gravitational well. As matter and energy approach the event horizon—the point of no return—they are subjected to increasingly intense gravitational forces, leading to phenomena such as time dilation and gravitational lensing.

Within the ergosphere—a region surrounding the event horizon—objects are dragged along the rotation of the black hole, experiencing frame-dragging effects. This phenomenon, predicted by Einstein's theory of general relativity, has profound implications for the behavior of matter and energy in the vicinity of black holes.

Magnetic Fields: Nature's Powerhouses

Black holes are surrounded by incredibly strong magnetic fields, generated by the rotation of the black hole itself and the accretion of matter from its surroundings. These magnetic fields can shape the behavior of charged particles in their vicinity, leading to the formation of astrophysical jets and accretion disks.

Accretion disks, composed of hot plasma swirling around the black hole, are threaded with magnetic field lines that channel matter towards the event horizon. As material spirals inward, friction within the disk generates intense heat and radiation, producing some of the brightest sources of X-rays and gamma-rays in the universe.

Astrophysical jets, narrow streams of high-energy particles, are launched from the vicinity of black holes along their spin axes. These jets can extend over vast distances, often spanning millions of light-years, and play a crucial role in shaping the cosmic landscape on both galactic and intergalactic scales.

Particle Acceleration: Nature's Particle Accelerators

Black holes are also powerful particle accelerators, capable of accelerating particles to velocities approaching the speed of light. Within the vicinity of black holes, particles can gain tremendous amounts of energy through a variety of mechanisms, including magnetic reconnection, shock acceleration, and interactions with intense electromagnetic fields.

These accelerated particles can emit radiation across the entire electromagnetic spectrum, from radio waves to gamma rays, providing astronomers with valuable insights into the underlying physical processes at work. By studying the high-energy emission from black holes, researchers can probe the extreme conditions near these cosmic behemoths and unravel the mysteries of particle acceleration in the universe.

Conclusion

The extreme environments around black holes represent some of the most fascinating and dynamic regions in the cosmos. From the intense gravity of the event horizon to the powerful magnetic fields shaping accretion disks and astrophysical jets, black holes exert a profound influence on their surroundings, driving processes that are fundamental to our understanding of the universe.

As we continue to study black holes and their environs with ever-improving observational and theoretical techniques, we gain valuable insights into the nature of gravity, magnetism, and particle physics in extreme conditions. By unraveling the mysteries of black holes, we not only expand our understanding of the universe but also glimpse the cosmic forces at play on the grandest scales imaginable.

Chapter 9

Theoretical Concepts

Introduction

Black holes, the enigmatic cosmic entities predicted by the theory of general relativity, have fascinated scientists and captivated the public imagination for decades. These gravitational monsters, born from the collapse of massive stars, possess such immense gravitational pull that not even light can escape their grasp, rendering them invisible to direct observation. Despite their mysterious nature, theoretical physics has provided us with intriguing concepts that offer insight into the behavior and properties of black holes. In this discourse, we delve into several theoretical concepts related to black holes, including wormholes, black hole thermodynamics, and the holographic principle, aiming to unravel the profound mysteries of these cosmic phenomena.

Wormholes: Gateways through Spacetime

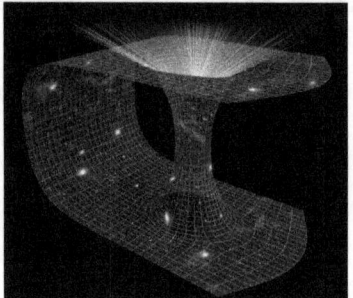

One of the most captivating theoretical concepts associated with black holes is the notion of wormholes. Proposed within the framework of general relativity, wormholes are hypothetical tunnels that connect two distinct regions of spacetime, potentially allowing for instantaneous travel between vast cosmic distances. Mathematically, wormholes are solutions to Einstein's field equations, exhibiting peculiar properties such as a throat region where spacetime is highly curved and a traversable conduit linking distant points.

Theoretical explorations of wormholes have fueled the imagination of scientists and science fiction enthusiasts alike, envisioning scenarios of interstellar travel and cosmic shortcuts. However, the feasibility of traversable wormholes remains highly speculative, primarily due to several daunting challenges. These include the destabilizing effects of exotic matter with negative energy density required to keep the wormhole open, as well as the intense gravitational forces near black holes, which could shred any object attempting to traverse the wormhole.

Nevertheless, the study of wormholes continues to inspire theoretical investigations into the nature of spacetime and the possibilities of traversing vast cosmic distances. Whether as hypothetical constructs for scientific inquiry or as narrative devices in popular culture, wormholes exemplify the boundless creativity and curiosity inherent in the human quest to comprehend the cosmos.

Black Hole Thermodynamics: Entropy and Information Paradoxes

In the realm of black hole physics, the marriage between thermodynamics and gravity has yielded profound insights into the nature of these cosmic behemoths. Black hole thermodynamics, a theoretical framework extending the laws of thermodynamics to black hole phenomena, has uncovered remarkable parallels between the behavior of black holes and thermodynamic systems.

Central to black hole thermodynamics is the concept of black hole entropy, which quantifies the microscopic disorder associated with the quantum states of matter engulfed by a black hole. Remarkably, the area of a black hole's event horizon serves as a measure of its entropy, leading to the formulation of the famous Bekenstein-Hawking entropy formula, which relates the entropy of a black hole to one-fourth of its surface area in Planck units.

The implications of black hole entropy extend beyond thermodynamics, touching upon profound questions in quantum mechanics and information theory. Perhaps most notably, the study of black hole entropy has fueled debates surrounding the information paradox, a conundrum stemming from the apparent loss of information when matter falls into a black hole and subsequently evaporates through Hawking radiation.

The resolution of the information paradox remains an active area of research, with proposals ranging from the conservation of information via subtle quantum correlations to radical revisions of our understanding of spacetime and quantum gravity. Regardless of the ultimate solution, the interplay between black hole thermodynamics and information theory underscores the deep connections between gravity, quantum mechanics, and the nature of reality.

The Holographic Principle: Information Encoding on Boundaries

A revolutionary theoretical concept that has emerged from the study of black holes is the holographic principle. Conceived initially in the context of black hole physics by Gerard 't Hooft and further developed by Leonard Susskind, the holographic principle posits that the information content of a spatial region can be fully encoded on its boundary surface.

At its core, the holographic principle challenges conventional notions of spatial locality and dimensional hierarchy, suggesting that a lower-dimensional surface can encode all the degrees of freedom and information content of a higher-dimensional volume. In the context of black holes, the holographic principle implies that the entropy and quantum states of a black hole can be entirely described by the degrees of freedom residing on its event horizon.

The profound implications of the holographic principle extend far beyond black hole physics, offering a new perspective on the nature of spacetime and the fundamental structure of reality. In particular, the holographic principle has found deep connections with the theory of quantum gravity, string theory, and the quest for a unified description of the fundamental forces of nature.

Moreover, the holographic principle has inspired groundbreaking developments in theoretical physics, such as the AdS/CFT correspondence, which establishes a duality between certain gravitational theories in anti-de Sitter space and conformal field theories living on the boundary of that space. This remarkable correspondence has provided novel insights into the nature of quantum gravity and has led to profound conjectures about the underlying principles governing the universe.

Conclusion

Theoretical concepts surrounding black holes, including wormholes, black hole thermodynamics, and the holographic principle, offer profound insights into the nature of gravity, spacetime, and the fundamental structure of the cosmos. While these concepts remain theoretical constructs awaiting empirical validation, they represent the forefront of human inquiry into the deepest mysteries of the universe.

As we continue to explore and refine our understanding of black holes and their associated phenomena, we embark on a journey of intellectual discovery that transcends the boundaries of current knowledge. Whether through mathematical formalism, thought experiments, or theoretical conjecture, the quest to unravel the mysteries of black holes illuminates the boundless potential of human curiosity and the enduring quest for cosmic understanding.

Chapter 10

Future Directions

Abstract:

Black hole astronomy stands at the forefront of modern astrophysics, offering tantalizing insights into the nature of spacetime, matter, and the universe itself. This paper delves into prospective avenues for research in black hole astronomy, focusing on the hunt for intermediate-mass black holes (IMBHs), the dynamics of black hole mergers, and their profound implications for fundamental physics. By examining the latest advancements and anticipated technological developments, we outline the potential trajectories of exploration and discovery in this captivating field.

Introduction:

The enigmatic nature of black holes continues to captivate astronomers and physicists alike, driving relentless pursuit towards understanding these cosmic phenomena. As observational capabilities expand and theoretical frameworks evolve, the study of black holes has entered an era of unprecedented growth. This paper delineates the future directions of black hole astronomy, highlighting key areas of inquiry such as the search for intermediate-mass black holes, the dynamics of black hole mergers, and the broader implications for fundamental physics.

Intermediate-Mass Black Holes:

Intermediate-mass black holes represent a crucial yet elusive class of cosmic entities, bridging the gap between stellar-mass black holes and supermassive black holes found at the centers of galaxies. The quest to detect and characterize IMBHs encompasses both observational and theoretical endeavors. Future astronomical surveys leveraging advanced instruments like the James Webb Space Telescope (JWST) and next-generation ground-based observatories promise to unveil the presence of IMBHs in diverse environments, from globular clusters to galactic nuclei. Furthermore, sophisticated numerical simulations and gravitational wave studies offer insights into the formation mechanisms and evolutionary pathways of IMBHs, shedding light on their role in galactic dynamics and cosmic evolution.

Black Hole Mergers:

The mergers of black hole binaries stand as cosmic spectacles, generating ripples in spacetime known as gravitational waves that propagate across the cosmos. Future observations by upgraded gravitational wave detectors such as LIGO (Laser Interferometer Gravitational-Wave Observatory) and its successors hold immense potential for probing the dynamics of black hole mergers with unprecedented precision. By analyzing the properties of merger events, including mass ratios, spin orientations, and electromagnetic counterparts, researchers aim to unravel the origins of binary systems and the underlying mechanisms driving their coalescence. Moreover, theoretical investigations into the gravitational wave signatures of extreme mass ratio inspirals (EMRIs) offer unique opportunities to probe the spacetime geometry near supermassive black holes, opening new windows into the realm of general relativity.

Implications for Fundamental Physics:

The study of black holes transcends the boundaries of astrophysics, intersecting with fundamental principles of physics such as quantum mechanics and gravity. Future research endeavors in this domain seek to reconcile the enigmatic properties of black holes with quantum field theory, ultimately unraveling the nature of spacetime at the most fundamental level. Quantum gravity theories, including string theory and loop quantum gravity, offer tantalizing frameworks for understanding the behavior of matter and energy in the vicinity of black holes. Additionally, investigations into the information paradox and black hole thermodynamics provide fertile ground for exploring the interplay between gravity, entropy, and quantum information.

The future of black hole astronomy brims with promise and excitement, fueled by technological innovation, theoretical ingenuity, and interdisciplinary collaboration. From the quest for intermediate-mass black holes to the dynamics of black hole mergers and the implications for fundamental physics, the trajectory of exploration in this field holds the potential to revolutionize our understanding of the cosmos. By charting a course towards new horizons, astronomers and physicists alike embark on a journey of discovery, guided by the gravitational whispers of these cosmic behemoths.

Conclusion

Conclusion

Throughout the journey we've embarked on in "Exploring the Mysteries of Black Holes," we've ventured into the depths of the universe, confronting some of its most enigmatic and beguiling phenomena. As we conclude our exploration, let's take a moment to reflect on the profound significance of black holes in shaping our understanding of the cosmos.

In our quest to comprehend the nature of black holes, we've encountered astonishing revelations about the fabric of spacetime, the laws of gravity, and the fate of matter under extreme conditions. From the elegant equations of general relativity to the awe-inspiring images captured by telescopes and detectors, our understanding of black holes has been refined and expanded over the decades, yet they remain as tantalizingly mysterious as ever.

One of the most remarkable aspects of black holes is their ability to warp space and time to an extraordinary degree, creating gravitational wells from which not even light can escape. These cosmic behemoths, born from the collapse of massive stars or forged in the crucible of galactic mergers, exert an irresistible pull on the fabric of the universe, shaping the evolution of galaxies and sculpting the cosmic landscape.

But black holes are not merely celestial curiosities—they are cosmic laboratories where the most extreme conditions imaginable come into play. From the searing temperatures of accretion disks to the mind-bending distortions of spacetime near the event horizon, black holes challenge our understanding of the laws of physics and beckon us to probe the limits of our knowledge.

Moreover, black holes serve as cosmic signposts, guiding us on a journey through the history of the universe. Their presence within galaxies provides crucial clues about the formation and evolution of cosmic structures, shedding light on the processes that have shaped the cosmos over billions of years.

Yet, amidst the awe and wonder inspired by black holes, profound questions remain unanswered. The nature of the singularity lurking at the heart of a black hole, the fate of information consumed by its voracious appetite, and the ultimate destiny of these cosmic leviathans continue to puzzle and intrigue astronomers and physicists alike.

Exploring the Mysteries of Black Holes | Amazir Amanath | 1st Edition

As we bring our exploration to a close, we are reminded of the boundless potential for discovery that black holes represent. They challenge us to push the boundaries of our knowledge, to seek answers to the most fundamental questions about the nature of reality, and to confront the mysteries that lie at the heart of the cosmos.

In the end, the study of black holes is not merely an academic pursuit—it is a profound and humbling endeavor that reminds us of our place in the vast tapestry of the universe. By unraveling the mysteries of black holes, we gain insight into the very fabric of reality itself, expanding our horizons and deepening our appreciation for the wonders that lie beyond the reach of our imagination.

As we turn our gaze once more to the heavens, let us continue our exploration with a sense of wonder and curiosity, knowing that the universe holds countless mysteries yet to be discovered, and that the journey of discovery is as infinite as the cosmos itself.

Amazir Amanath
(Diploma in Astronomy)

Exploring the Mysteries of Black Holes" takes readers on an exhilarating journey through the cosmos, delving into one of the most captivating and enigmatic phenomena in the universe. Written by renowned astrophysicist Amazir Amanath, this book offers a comprehensive and accessible exploration of black holes, from their awe-inspiring origins to their profound implications for our understanding of the cosmos.

www.ingramcontent.com/pod-product-compliance
Lightning Source LLC
Chambersburg PA
CBHW070949220526
45471CB00007B/2951